SOLUTIONS MANUAL

WILLIAM STALLINGS • RICHARD VAN SLYKE • STAN SCHATT

BUSINESS DATA COMMUNICATIONS

THIRD EDITION

WILLIAM STALLINGS • RICHARD VAN SLYKE

PRENTICE HALL Upper Saddle River, NJ 07458

Acquisitions Editor: ***Alan Apt***
Production Editor: ***James Buckley***
Supplement Editor : ***Laura Steele***
Production Coordinator: ***Donna Sullivan***
Cover Manager: ***Paul Gourhan***
Cover Designer: ***Liz Nemeth***

Simon & Schuster/A Viacom Company
Upper Saddle River, NJ 07458

Printed in the United States of America

10 9 8 7 6 5 4 3 2 1

ISBN 0-13-750994-4

Prentice-Hall International (UK) Limited, *London*
Prentice-Hall of Australia Pty. Limited, *Sydney*
Prentice-Hall Canada, Inc., *Toronto*
Prentice-Hall Hispanoamericana, S.A., *Mexico*
Prentice-Hall of India Private Limited, *New Delhi*
Prentice-Hall of Japan, Inc., *Tokyo*
Simon & Schuster Asia Pte. Ltd., *Singapore*
Editora Prentice-Hall do Brasil, Ltda., *Rio de Janeiro*

CONTENTS

CHAPTER 1

INTRODUCTION

ANSWERS TO QUESTIONS:

1. Communication technology can make geographically dispersed companies more manageable; they can help top-heavy companies trim down middle management; and they can help companies break down barriers between different divisions.

2. Voice communications, data communications, image communications, and video communications are all found on networks.

3. Compact disk technology has resulted in the development of optical disks that permit massive storage of information including images as well as text.

4. Today the options for selecting voice and data equipment are far more complex than in previous years. Managers must face the prospect that since equipment is cheaper than in the past, there is a movement toward decentralization which makes coordination and management of resources much more difficult.

5. Optical fiber transmission has become more popular in the past few years because of its high capacity and security characteristics.

6. TCP/IP, ISO, and SNA are three approaches to organizing communications software.

7. Multiplexing and data compression are two approaches to increasing the efficiency of data transmission.

8. Application software performs a specific function such as accounting while interconnection software ensures that all computers and terminals speak the same language and can be connected together.

CHAPTER 2

BUSINESS INFORMATION

ANSWERS TO PROBLEMS:

1. a) 5 bits ($2^5 = 32$)

   b) 4 bits ($2^4 = 16$, this is one reason why hexadecimal is used)

   c) 17 bits (The size of the set is 60 x 60 x 24 = 86,400
   $2^{17} = 2 \times 65,536 = 131,072$)

   d) 28 bits ($2^{10} \times 2^{10} \times 2^8 = 2^{28} = 1024 \times 1024 \times 256 = 268,435,456$)

   e) 33 bits ($2^{10} \times 2^{10} \times 2^{10} \times 2^3 = 2^{32} \times 2 = 4.3 \times 10^9$
   Thus, for example, every person in the world can be individually identified with a string of 33 "1"'s and "0"'s.)

2. a) 4096 levels

   b) 2048 x 2048 x 12 = 50,331,648 bits

   c) (5 x 50,331,648)/1,544,000 = 163 seconds = 2 minutes 43 seconds

   d) (5 x 50,331,648)/2 = 126 Mbps

   e) 4 times 126 Mbps = 503 Mbps.

3. a) $600 \times 2^{20} = 600 \times 1,048,576 = 629,145,600$ bytes
   = 5,033,164,800 bits ≅ 5 Gb.

   b) Data (text):
   1000 pages x (9 x 6) lines x (6.5 x 10) char/line x 8 bits/char
   = 28.1 Mb
   Images:
   100 x (1024 x 768 x 8) = 629.1 Mb
   Audio:
   30 minutes x 60 seconds/minute x 16,000 samples/second x 6 bits/sample
   = 172,800,000 = 172.8 Mb
   The total is 830 Mb which corresponds to 5,000/830 ≅ 6 volumes per CD.

   c) 830 Mb/1.544 Mbps = 537.6 seconds ≅ 9 minutes

d) Data: 28.1/3 = 9.4 Mb
Image: 629.1/10 = 62.9 Mb
Audio: 172.8/2 = 86.4 Mb
Total: 158.7 Mb or about 32 volumes on a CD.
The transmission time would be 158.7/1.544 = 103 seconds = 1 minute, 43 seconds.

4. a) We allow one extra character per word for spaces and punctuation. Thus, the total number of characters is about 6.1 x 44,000,000 = 268,400,000 bytes, or 2.147 Gb.

b) Transmission time at 1.544 Mbps is 2,147/1.544 = 1391 seconds or about 23 minutes. At 2.488 Gbps the transmission time is under a second.

c) The text (note we have not considered the pictures) would take less than ½ of a CD (2.147/5 = 0.43).

5. a) 25 x 512 x 512 x 8 = 52.4 Mb

b) 30 x 512 x 512 x 8 = 62.9 Mb

c) 25 x 30 x 512 x 512 x 8 = 1.6 Gb for each study. A CD can hold 5/1.6 = 3.1 studies.

CHAPTER 3

DISTRIBUTED PROCESSING

ANSWERS TO QUESTIONS:

1. In a fully centralized data processing facility, the computers, application programs, data, support staff, and network equipment are all found at that site.

2. A centralized data processing facility often offers economies of scale in the purchase and operation of equipment and software. It can also afford professional programmers on staff, provide standards for equipment and programming, and provide adequate security.

3. A distributed data processing strategy is one in which computers are dispersed throughout an organization based on maximum effectiveness.

4. Distributed small systems can result in a loss of centralized control and may result in incompatible systems being purchased.

5. Connecting distributed data processing systems can result in the sharing of expensive resources and the exchanging of data. It also increases reliability since the impact of systems failures is reduced.

6. Horizontal partitioning of applications means one application replicated on a number of machines or a number of different applications distributed among a number of machines. Vertical partitioning means one application split up into components that are dispersed among a number of machines.

7. A distributed database offers a number of advantages including the ability to have a distributed design that reflects the company's structure and the ability to store data locally, minimizing the effect of a breakdown at the computer site. By having data distributed smaller machines may be utilized.

8. Database objectives include accessibility of data and security and privacy as well as the integrity of data. Communications objectives are to minimize the communications load and delays imposed by the use of a communications facility.

9. A factory can contain a number of programmable controllers as well as robots that contribute to the automation of the manufacturing process.

10. Distributed data processing generates requirements for components and networking which include connectivity, availability, and performance.

11. The key features of an intranet are the use of Internet-based standards, such as HTML,

and SMTP; the use of TCP/IP for local and wide-area networking; content inaccessible to the general Internet. The intranet is managed by the organization while the Internet is not.

ANSWERS TO PROBLEMS:

1. $n(n-1)/2$, 1225.

CHAPTER 4

TRANSMISSION AND TRANSMISSION MEDIA

ANSWERS TO QUESTIONS:

1. A continuous or analog signal is one in which the signal intensity varies in a smooth fashion over time while a discrete or digital signal is one in which the signal intensity maintains one of a finite number of constant levels for some period of time and then changes to another constant level.

2. Amplitude, frequency, and phase are three important characteristics of a periodic signal.

3. The spectrum of a signal is the frequencies it contains while the bandwidth of a signal is the width of the spectrum.

4. Since, when the frequency components outside of the 300 - 3400 Hz range are subtracted, the remaining voice signal still sounds acceptable, the telephone networks can save in bandwidth and therefore money by using only these frequencies.

5. Attenuation is the gradual weakening of a signal over distance.

6. Because of delay distortion, various components of a signal arrive at a receiver at different times. At very high speeds, it is possible for some of the signal energy in one bit position to spill over into other bit positions.

7. Noise uniformly distributed across the frequency spectrum is known as white noise.

8. With guided media, the electromagnetic waves are guided along an enclosed physical path whereas unguided media provide a means for transmitting electromagnetic waves but do not guide them.

9. The twisting of the individual pairs reduces electromagnetic interference. For example, it reduces crosstalk between wire pairs bundled into a cable.

10. Twisted pair wire is subject to interference, limited in distance, band width, and data rate.

11. Optical fiber consists of a column of glass or plastic surrounded by an opaque outer jacket. The glass or plastic itself consists of two concentric columns. The inner column called the core has a higher index of refraction than the outer column called the cladding.

12. Point-to-point microwave transmission has a high data rate and less attenuation than twisted pair or coaxial cable. It is affected by rainfall, however, especially above 10 GHz. It is also requires line of sight and is subject to interference from other microwave

transmission, which can be intense in some places.

13. Direct broadcast transmission is a technique in which satellite video signals are transmitted directly to the home for continuous operation.

14. A satellite must use different uplink and downlink frequencies for continuous operation in order to avoid interference.

15. Broadcast is omnidirectional, does not require dish shaped antennas, and the antennas do not have to be rigidly mounted in precise alignment.

16. While fiber optics and satellite transmission both have advantages and disadvantages, the bandwidth advantage of optical fiber is helping it prevail and dominate the market except for special applications.

ANSWERS TO PROBLEMS:

1. 5
   7
   6

2.

a)			b)		
	D=	0010001		D=	00100011
	d=	0010011		d=	00100110
	H=	0001001		H=	00010011
	h=	0001011		h=	00010110

All represented with the least significant bit first.

3. a) $\sin(2\pi ft - \pi) + \sin(2\pi ft + \pi) = 2\sin(2\pi ft + \pi)$ or $2\sin(2\pi ft - \pi)$ or $-2\sin 2\pi ft$

   b) $\sin 2\pi ft + \sin(2\pi ft - \pi) = 0$.

4. Modems for use on the telephone network must work within the < 4 kHz bandwidth of a voice channel. The figure quoted in the text assumes one can use the full bandwidth of a twisted pair which is considerably larger.

5. The wavelength is 10,000 km. Half is 5,000 km which is about 3107 miles which is comparable to the east to west dimension of the continental US. While an antenna of this size is impracticable, the U.S. Defense Department was trying to take over large parts of Wisconsin and Michigan to make an antenna many miles in dimension. One proposal was to cover 186,000 sq. mi (equivalent to a square of side 431 mi.).

6. a) The length of a ½ wavelength antenna is given by: $\lambda = c/f$ = (300,000 km./sec)/300Hz = 1,000 km.; $\lambda/2$ = 500 km.

   b) The carrier frequency corresponding to $\lambda/2$ = 1 m. is given by: λ = 2 m. $f = c/\lambda$ = (300,000,000 m/sec)/2m = 150 MHZ.

7. There is a typo in this problem. The **solid** curve is $\sin 2\pi t$. The **dotted** curve is $2\sin(4\pi t + \pi)$.

8. Shannon's formula implies: $\log_2(1+S/N)$ = 28,800/3,300 = 8.7273. $2^{8.7273}$ = 423.81. Thus, 422.81 = S/N; in decibels, this is 10 log(422.81) = 26.26.

9. $c = \lambda f$, $\lambda = 2 \times 2.5\text{x}10^{-3} = 5\text{x}10^{-3}$; $f = c/\lambda = 30 \times 10^7/5 \times 10^{-3} = 6 \times 10^{10}$ = 60 GHz.

10. The received signal is, essentially, the same. The received power will increase by a factor of 4.

CHAPTER 5

COMMUNICATION TECHNIQUES

ANSWERS TO QUESTIONS:

1. A modem converts digital information into an analog signal, and conversely.

2. Cost, capacity utilization, and security and privacy are three major advantages enjoyed by digital transmission over analog transmission.

3. With amplitude-shift keying, binary values are represented by two different amplitudes of carrier frequencies. This approach is susceptible to sudden gain changes and is rather inefficient.

4. The three basic forms of modulation of analog signals for digital data are amplitude shift keying, frequency shift keying, and phase shift keying.

5. Non return-to-zero-level (NRZ-L) is a data encoding scheme in which a negative voltage is used to represent binary one and a positive voltage is used to represent binary zero. A disadvantage of NRZ transmission is that it is difficult to determine where one bit ends and the next bit begins.

6. The beginning of a character is signaled by a start bit but with a value of binary zero. A stop (binary one) follows the character.

7. Asynchronous transmission requires an overhead of two or three bits per character, and is, therefore, significantly less efficient than synchronous transmission.

8. Data circuit-terminating equipment (DCE) mediates between a user's data terminal equipment (DTE) and a network or transmission facility.

9. Printers normally have a buffer. They receive information from a computer, and then they send an acknowledgment signal to the computer when they are ready to have more information sent.

10. The sliding window flow control technique can send multiple frames before waiting for an acknowledgment. The stop & wait approach requires acknowledgments after each frame.

11. Error control consists of error detection and error correction.

12. Go-back-N ARQ is a form of error control in which a destination station sends a negative acknowledgment (NAK) when it receives an error. The source station receiving the NAK

will retransmit the frame in error plus all succeeding frames transmitted in the interim.

ANSWERS TO PROBLEMS:

1. a) Each character has 25% overhead. For 10,000 characters there are 20,000 extra bits. This would take an extra 20,000/2,400 = 8 1/3 seconds.

   b) The file takes 10 frames or 480 additional bits. The transmission time for the additional bits is 480/2400 = 0.2 seconds.

   c) Ten times as many extra bits and ten times as long for both a) and b).

   d) The number of overhead bits would be exactly the same, and the time would be decreased by a factor of 4 = 9600/2400.

2. For each case, compute the fraction g of transmitted bits that are data bits. Then the maximum effective data rate R is:

   $R=gB$

   (a) There are 7 data bits, 1 start bit, 1.5 stop bits, and 1 parity bit

   $g = 7/(1 + 7 + 1 + 1.5) = 7/10.5$
   $R = 0.67B$

   (b) Each frame contains 48 + 128 = 176 bits. The number of characters is 128/8 = 16, and the number of data bits is 16 x 7= 112.

   $R = (112/176)B = 0.64B$

   (c) Each frame contains 48 + 1024 = 1072 bits. The number of characters is 1024/8 = 128, and the number of data bits is 128 x 7 = 896.

   $R = (896/1072)B = 0.84B$

3. On Link A-B the packet transmission time is T_t = 1000 bits/100,000 bps = 0.01 seconds. The propagation time is T_p = 10 μsec/mile x 2,000 miles = 0.02 seconds. After the beginning of transmission of a frame it takes 0.02 for the first transmitted bit to get to B, 0.01 seconds for the rest of the frame to arrive, and a final 0.02 seconds for the acknowledgment to be received. With a sliding window with window size 3, 3 packets can be sent, but no more, before the acknowledgment is received. Thus we can send three frames every 0.05 seconds, for an effective data rate of 3000/0.05 = 60,000 bps.

Therefore the link B-C must have a high enough data rate to keep up.

Under stop and wait flow control, one frame can be sent for each cycle which consists of one propagation time (in this case 10 μsec/mile x 500 miles = 0.005 seconds), one transmission time, and another propagation time. Let R be the data rate of B-C. Then the effective utilization U of the channel is:

U = 1000/(1000 + 2 x 0.005 x R), and we must have UR ≥ 60,000. Solving, we obtain R ≥ 150 Kbps.

4. a) The speech is 23 x 60 = 1,380 seconds. The sampling rate has to be at least 44,000 samples per second, for a total of 1,380 x 44,000 = 60,720,000 samples. To get the number of bits to discretize each sample we note that 2^{16} = 65,536 which is about 64,000. So the total number of bits is 16 x 60,720,000 = 971,520,000 bits or 121,440,000 bytes.

   b) 4 pages x 7 3/8" = 29 ½ linear inches. The total number of lines is 29 ½ x 5 x 1/3 = (59/2)x(16/3) = 472/3. There are 77 characters per line. So the total number of characters = total number of bytes is 472x77/3 = 12,115.

   c) The total area to be scanned is 4.5 x 29.5 = 132.75 square inches. The number of pixels per square inch is 1200 x 1200 = 1,440,000. The total number of bits is 1,440,000 x 132.75 = 191,160,000 bits and the total number of bytes is 23,895,000.

5. a) The propagation delay is 936,237,665.3592 km./ 300,000 km/sec = 3121 seconds, which is almost an hour.

   b) The shortest time after the message is sent in which an acknowledgment can be received back from the Galileo is 2 x 3121 = 6242 seconds. The time for transmitting a packet is 2000x8/134,400 = 0.119. Thus W = 6242/0.119 = 52,454 packets can be transmitted before an acknowledgment arrives.

   c) The number of bits required is $\log_2 52{,}454$ rounded up or 16 bits.

   d) The frame transmission time = 2000x8/8 = 2,000 seconds. W = 6242/2000, rounded up to 4. W=4 can be supported with the standard HDLC frames.

CHAPTER 6

TRANSMISSION EFFICIENCY

ANSWERS TO QUESTIONS:

1. Multiplexing is cost-effective because the higher the data rate, the more cost-effective the transmission facility.

2. Interference is avoided under frequency division multiplexing by the use of guard bands, which are unused portions of the frequency spectrum between subchannels.

3. A synchronous time division multiplexer interleaves bits from each signal and takes turns transmitting bits from each of the signals in a round-robin fashion.

4. T-1 lines are used for such applications as private voice and data networks, video teleconferencing, and high-speed digital facsimile information.

5. The use of private T-1 lines is attractive to companies because not only is it cost-effective but also it permits simpler configurations than the use of a mix of lower speed offerings.

6. A statistical time division multiplexer is more efficient than a synchronous time division multiplexer because it allocates time slots dynamically on demand and does not dedicate channel capacity to inactive low speed lines.

7. Run length encoding is a data compression technique that consists of replacing sequences of repeating characters with a short code.

8. The basic difference between North American and international TDM carrier standards is that the North American DS-1 carrier has 24 channels while the international standard is 30 channels. This explains the basic difference between the 1.544 Mbps North American standard and the 2.048 Mbps international standard.

9. As load increases, the buffer size and delay increase until the load approximates the capacity of the shared channel when both become infinite.

10. Lossless compression is used when the reconstituted information must be absolutely identical with the original. Data information such as text or data usually requires lossless compression. The large memory and communication requirements of image, audio, and video often make it necessary to achieve the higher compression factors possible with lossy compression where there is allowed "perceptually insignificant" differences between the original information and the reconstituted information.

ANSWERS TO PROBLEMS:

1. One of the 24 voice channels of the DS-1 format is used in digital transmission for synchronization, leaving 23 channels. One bit per for each channel's byte is used for control information. So each of the remaining 23 channels gets 7 bits 8,000 times a second (the voice sampling rate) for a channel rate of 56 kbps.

2. a) You may use either 64 kbps or 56 kbps for PCM. I will use 64 kbps. n = 64000 x 3 x 60 = 11,520,000 bits.

   b) One page of text has 65 x 8 x 55 = 28,600 bits. The number of pages that corresponds to one 3 minute call is 11,520,000/28,600 = 402.8, a good sized book.

   c) One faxed page corresponds to 8 x 200 x 10.5 x 100 / 10 = 168,000 bits. Thus 11,520,000/168,000 = 68.6 faxed pages to one phone call.

3. a) 8 x 10 x 1200^2 x 10 = 1,152,000,000 bits = 144,000,000 bytes.

   b) 2^{30} is a little over 1 billion (1,073,741,000). Thus, n=10 for each color.

4. a) $7 \times 11 = 77$ in^2; 2400×2400 pixels/ in^2; bits = 443,520,000; bytes = 55,440,000.

   b) time = 443,520,000/19,200 = 23,100 seconds = 6.42 hours.

5. Voice sampling rate = 2 x 4 kHz = 8 kHz; 6 bits/sample

Thus:			
30 voice channels:		30 x 8 x 6	= 1440 kbps
	1 synchronous bit/channel:	30 x 8	= 240 kbps
	1 synchronous bit/frame:	1 x 8	= 8 kbps
TOTAL			1688 kbps

CHAPTER 7

WIDE-AREA NETWORKS

ANSWERS TO QUESTIONS:

1. Wide area networks (WANs) are used to connect stations over very large areas that may even be worldwide while local area networks (LANs) connect stations within a single building or cluster of buildings. Ordinarily, the network assets supporting a LAN belong to the organization using the LAN while in WANs, network assets of service providers are often used. LANs also generally support higher data rates than WANs.

2. It is advantageous to have more than one possible path through a network for each pair of stations to enhance reliability in case a particular path fails.

3. The control unit of a circuit-switching device establishes connections, maintains connections, and tears down connections.

4. A blocking network is one in which two stations may be unable to be connected depending on traffic, while a nonblocking network permits any pair of stations to be connected at any time.

5. A busy hour is the average traffic load expected over the course of the busiest hour of use during the course of a day.

6. Private networks offer users potential cost savings plus a greater degree of network control.

7. It is not efficient to use a circuit switched network for data since much of the time a typical terminal-to-host data communication line will be idle. Secondly, the connections provide for transactions at a constant data rate which limits the utility of the network in interconnecting a variety of host computers and terminals.

8. A value added network (VAN) is a public packet switched network in which the network provider owns a set of packet switching nodes and links these together with leased lines provided by a carrier such as AT&T.

9. Packet switching is impractical for digitized voice transmission because in packet switching delay is significant and varies. Voice traffic will not sound normal.

ANSWERS TO PROBLEMS:

1. We make the following assumptions:
   - -- the packet size P includes the header

-- an acknowledgment has size H for both virtual circuits and circuit switching
-- if a packet is partially filled it is reduced in length to less than P

We break the end-to-end delay for circuit switching, virtual circuits, and datagrams into four parts: setup, time for first bit to arrive at destination, time for the remaining transmitted bits to arrive, and time for acknowledgment to reach the originator.

(a)

	Circuit Switching	Virtual Circuit	Datagram
Setup	S	S	0
First bit arrived	ND	(N-1)P/B + ND	(N-1)P/B + ND
Additional time to last bit arrived	L/B	$\frac{L+\lceil L/(P-H)\rceil H}{B}$	$\frac{L+\lceil L/(P-H)\rceil H}{B}$
Acknowledgment	ND + H/B	ND + NH/B	0
GIVEN VALUES SUBSTITUTED: S=0.2, N=4, D=0.001, H=16, P=1024, L=3200, B=9600			
Setup	0.2	0.2	0
First bit arrived	0.004	0.324	0.324
Additional time to last bit arrived	0.33333	0.34	0.34
Acknowledgment	0.00567	0.01067	0.0
Total	0.54300	0.87467	0.66400

(b) From the formulas given above we note that for any positive values of the parameters, the delay for datagrams is less than for virtual circuits. similarly, circuit switching always causes less delay than virtual circuits. So the only interesting comparison is circuit switching versus datagrams.

To calculate the break-even point we consider the difference between the circuit switched delay T_{cs} and the datagram delay T_{dg}. This is given by:

$$T_{cs}-T_{dg}=[S+2ND+\frac{(H+L)}{B}]-[ND+\frac{(N-1)P}{B}+\frac{L+\lceil L/(P-H)\rceil H}{B}]$$
$$=[S+ND+\frac{H}{B}]-[\frac{(N-1)P+\lceil L/(P-H)\rceil H}{B}$$

The tradeoff is the setup time plus ND + H/B (the last two deal with acknowledgments) against the time it takes for the first packet to begin to reach the destination plus the time taken in transmitting overhead.

2. a) Each connection takes 2 hops. Hence the average path length is 2.

   b) In a loop we assume each pair takes the shortest path. By symmetry, the average number of hops from a given fixed node averaged over the shortest path lengths to its possible destinations is the same for each node. This same average is then the network average. Remember that 1+2+3+...+k = k(k+1)/2. We have two cases depending on whether N is odd or even. In both cases we number the nodes 0, 1, ... , N-1 and consider the average path length from node 0 to all other nodes. If N is even then Node 1 is one hop away, Node 2 is 2 hops and (N-1)/2 is (N-1)/2 hops away. Node (N+1)/2 is also (N-1)/2 hops away, Node (N+3)/2 is (N-3)/2 hops away on down to Node N which is 1 away. The sum of the path lengths is 2 x (1 + 2 + ... + (N-1)/2) = ((N-1)/2) x ((N+1)/2) and the average is (N+1)/4. A similar argument for N even leads to an average of $(N/2)^2/(N-1)$. In both cases the average is about N/4.

   c) One hop.

3. The Layer 2 flow control mechanism regulates the total flow of data between DTE and DCE. Either can prevent the other from overwhelming it. The Layer 3 flow control mechanism regulates the flow over a single virtual circuit. Thus, resources that are dedicated to a particular virtual circuit can be protected from overflow in either the DTE or DCE.

4. On each end, a virtual circuit number is chosen from the pool of locally available numbers and has only local significance. Otherwise, there would have to be global management of numbers.

CHAPTER 8

HIGH-SPEED WIDE-AREA NETWORKING

ANSWERS TO QUESTIONS:

1. Frame relay, ATM, SMDS, and BISDN; although their near term commercial availability is not universal, especially for ATM and BISDN.

2. Frame relay is much simplified, compared to X.25. The major differences are that frame relay uses out-of-channel signaling while X.25 uses all in-channel control. In frame relay there is no "hop-by-hop" flow control or error control as there is in X.25. If a frame error is detected it is just dropped rather than being retransmitted. Similarly, on an end-to-end basis, there is no error control or flow control except what is provided by higher level protocols outside of frame relay. Finally, frame relay is a two level (physical and link) layer protocol and the multiplexing of logical channels takes place at Level 2 rather than in the Level 3 Packet Layer as in X.25.

3. Because frame relay is so much simpler than X.25, the processing to support switching can be reduced and higher data rates than for X.25, up to several megabits per second, can be supported. On the other hand, because of the lack of hop-by-hop flow control, the user of frame relay has fewer tools to manage network congestion. The effective use of frame relay also depends on the channels being relatively error free. For example, this is true for fiber optics, but probably not for most forms of broadcast, wireless transmission.

4. Congestion is controlled in frame relay in two ways. The first, **congestion avoidance**, is used at the onset of congestion to minimize its effect on the network. The signaling methods in this case are **explicit** and make use of the backward explicit congestion modification (BECN) and forward explicit congestion notification (FECN) bits of the frame relay header. These bits inform users that they may encounter congestion in their use of the network. The other approach, **congestion recovery**, is used to maintain network integrity in the face of severe congestion. The principle mechanism is the dropping of packets. This signals **implicitly** to the user's higher protocol levels of congestion. The user or the network may set the **discard eligibility bit** of the header to indicate frames which could be discarded in preference to other frames with this bit not set.

5. The most obvious feature of ATM compared to frame relay is that ATM makes use of a 53 byte fixed length cell while the frame in frame relay is much longer, and may vary in length, both in its header and its data fields. Additionally, error checking is only done on the header in ATM rather than on the whole cell or frame. Virtual channels of ATM that follow the same route through the network are bundled into paths. A similar mechanism is not used in frame relay.

6. A **virtual channel** is a logical connection similar to virtual circuit in X.25 or a logical channel in frame relay. In ATM, virtual channels which have the same endpoints can be grouped into **virtual paths**. All the circuits in virtual paths are switched together; this offers increased efficiency, architectural simplicity, and the ability to offer enhanced network services.

7. SMDS is connectionless which offers advantages in applications characterized by isolated short transactions. However, data elements can arrive out of order in SMDS which is not supposed to occur in ATM and frame relay. SMDS also has several useful network management features including address-screening and user statistics which are not available in many ATM and frame relay implementations.

8. **Broadband Integrated Services Data Network (BISDN)** will provide interactive (conversational, messaging, retrieval) services and distribution (broadcast, and individual) services on a worldwide basis using modern broadband facilities. It extends ISDN, and is built on, and uses ATM for its lower layers.

ANSWERS TO PROBLEMS:

1. (a) We reason as follows. A total of X octets are to be transmitted. This will require a total of $\lceil X/L \rceil$ cells. Each cell consists of (L + H) octets. Thus

$$N = \frac{X}{\lceil X/L \rceil (L+H)}$$

The efficiency is optimal for all values of X which are integer multiples of the cell information size. In the optimal case, the efficiency becomes

$$N_{opt} = \frac{L}{L+H}$$

For the case of ATM, with L = 48 and H = 5, we have Nopt = 0.91

(b) Assume that the entire X octets to be transmitted can fit into a single variable-length cell. Then

$N = X/(X + H + H_v)$

(c) N for fixed-sized cells has a sawtooth shape. For long messages, the optimal achievable efficiency, N_{opt}, is approached. It is only for very short cells that efficiency is rather low. For variable-length cells, efficiency can be quite high,

efficiency is rather low. For variable-length cells, efficiency can be quite high, approaching 100% for large X. However, it does not provide significant gains over fixed-length cells for most values of X.

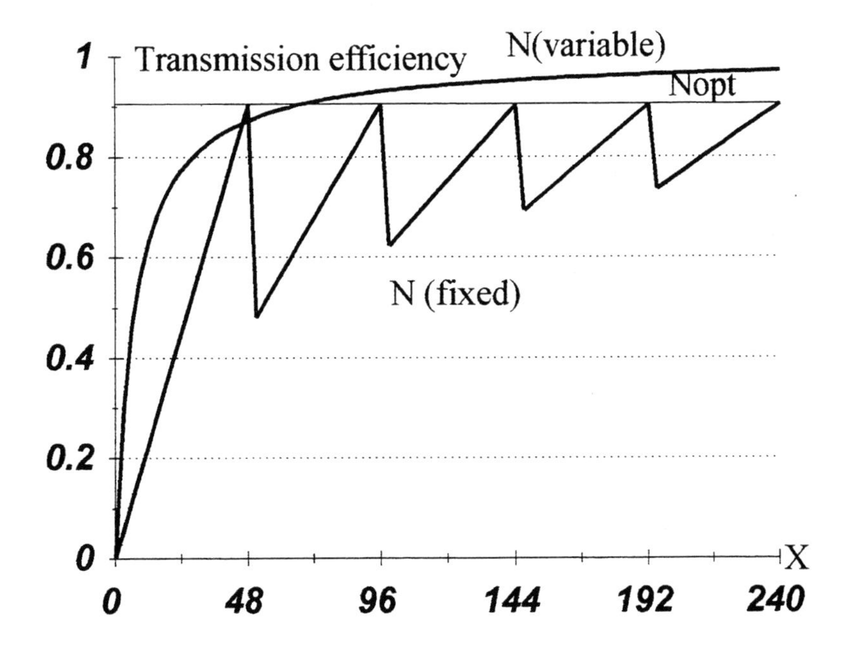

2. (a) As we have already seen in Problem 1:

$$N = L/(L + H)$$

(b) $D = (8 \times L)/R$

(c) A data field of 48 octets, which is what is used in ATM, seems to provide a reasonably good tradeoff between the requirements of low delay and high efficiency.

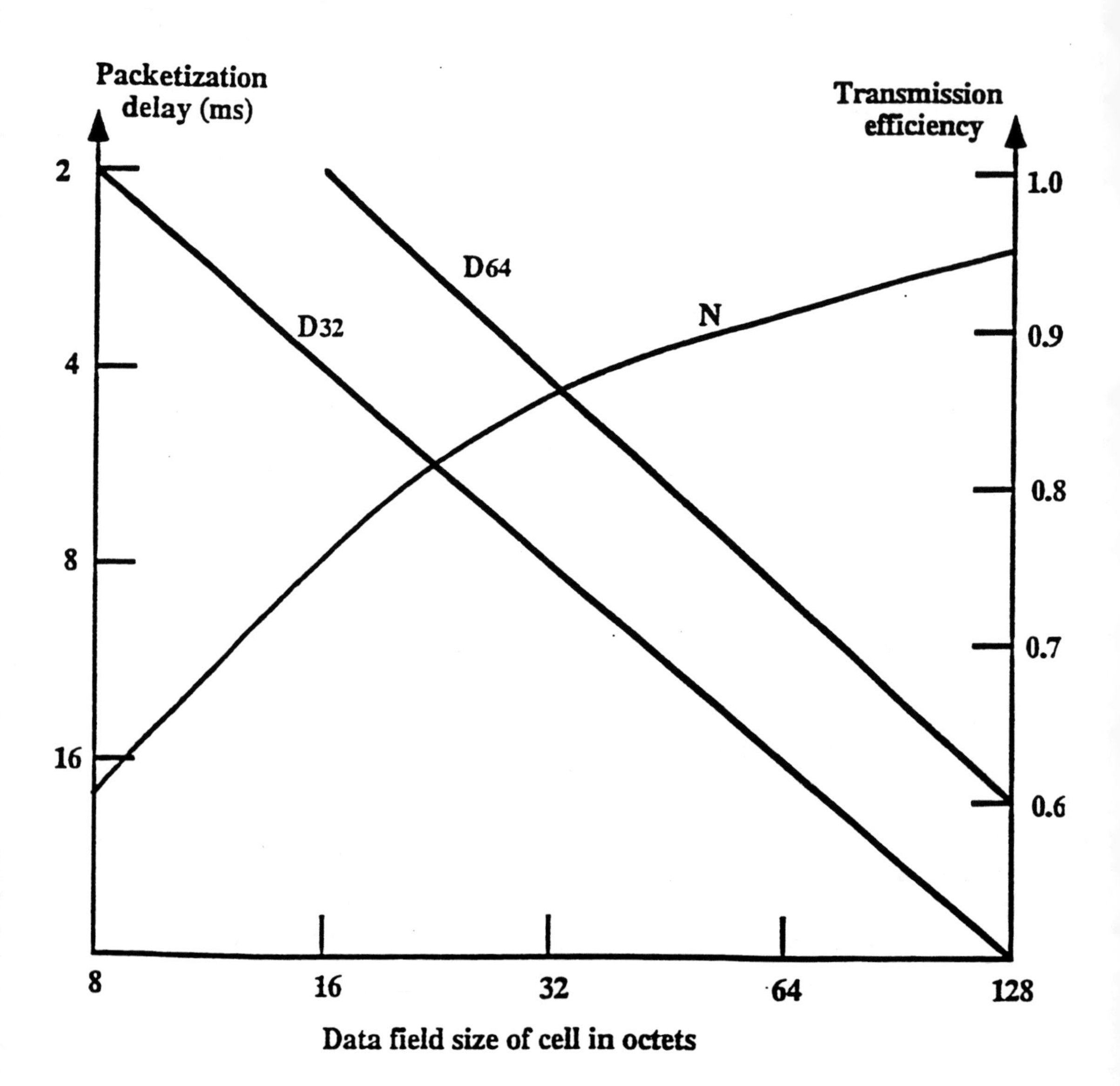
Packetization delay (ms)
Transmission efficiency
2
4
8
16
1.0
0.9
0.8
0.7
0.6
D64
D32
N
8
16
32
64
128
Data field size of cell in octets

3. a) For this application, you probably would prefer a connectionless (datagram) service to avoid going through an extensive virtual set up to send a small amount of information. SMDS is therefore attractive. In addition, one essentially pays for a single connection for each site, rather than multiple connections as in a dedicated network. The short delay requirement would seem to rule out X.25 service, as well.

b) In this case, the error rates on some of the channels could be relatively high. Therefore the hop-by-hop error control typical of X.25 packet switching is attractive. Since the data rates are modest, there is no need to give up the error and flow control advantages of X.25 for the other services.

c) Here, in order to handle the real time requirements with other data, you need to be able to interrupt data traffic with real time traffic. The delay in doing this is determined by the length of the data unit. The short 53 byte length of an ATM cell makes ATM attractive for this application. The virtual path concept of ATM also helps support the large number of processes by setting up a small number of paths between the few locations.

4. a) The transmission time for one cell through one switch is $t = (53 \times 8)/(43 \times 10^6) = 9.86\mu s$.

b) The maximum time from when a typical video cell arrives at the first switch (and possibly waits) until it is finished being transmitted by the 5th and last one is $2 \times 5 \times 9.86\mu s = 98.6\mu s$.

c) The average time from the input of the first switch to clearing the fifth is $(5 + 0.6 \times 5 \times 0.5) \times 9.86\mu s = 64.09\mu s$.

d) The transmission time is always incurred so the jitter is due only to the waiting for switches to clear. In the first case the maximum jitter is $49.3\mu s$. In the second case the average jitter is $64.09 - 49.3 = 14.79\mu s$.

CHAPTER 9

LOCAL NETWORKS

ANSWERS TO QUESTIONS:

1. Computer room networks require very high data rates and usually are concerned with transfer of large blocks of data.

2. Personal computers are used by individual professionals and secretarial personnel as well as networked for modest application loads. Minicomputers provide departmental time sharing using commonly used applications and thus generate heavier traffic than PCS. Mainframes are used for large database and scientific applications.

3. Network topology refers to the way in which the end parts or stations attached to the network are interconnected.

4. Baseband LAN systems can only extend about 1 km without repeaters because of attenuation of their signals.

5. On a dual-cable configuration, the inbound and outbound paths are separate cables with the head end simply a passive connector between the two while on a split configuration, the inbound and outbound paths are different frequencies on the same cable. A frequency converter at the head end is used to convert between the inbound and outbound frequencies.

6. By allowing subscriber locations to connect to other subscriber locutions, a linear strategy attempts to minimize the amount of cable to be laid to all subscriber locations. A star strategy uses a separate cable from each subscriber location to a concentration point.

7. Carrier sense multiple access with collision detection (CSMA/CD) is a form of medium access control in which a station listens to the medium to try to see if another transmission is in progress. If the medium appears idle, the station transmits. If a collision occurs, the workstations wait a random amount of time before trying again.

8. Under heavy loads, CSMA/CD's performance declines because of the number of collisions. The token bus functions more efficiently under a heavy load.

ANSWERS TO PROBLEMS:

1. HDLC has only one address field. In a local network any station may transmit to any other station. The receiving station needs to see its own address in order to know that the

data is intended for itself. It also needs the sending address in order to reply. HDLC could be used on a start topology because the circuit is transparent.

2. (a) $T = (8 \times 10^6 \text{ bits})/(64 \times 10^3 \text{ bps}) = 125$ seconds

(b) Transfer consists of a sequence of cycles. One cycle consists of:

o Data Packet = Data Packet Transmission Time + Propagation Time
o ACK Packet = ACK Packet Transmission Time + Propagation Time

Define:

C	=	Cycle time
Q	=	data bits per packet
T	=	total time required
T_d	=	Data Packet Transmit Time
T_a	=	ACK Packet Transmit Time
T_P	=	Propagation Time

Then

T_P	=	$(D)/(200 \times 10^6 \text{ m/sec})$
T_a	=	$(89 \text{ bits})/(B \text{ bps})$
T_d	=	$(P \text{ bits})/(B \text{ bps})$
T	=	$(8 \times 10^6 \text{ bits} \times C \text{ sec/cycle})/(Q \text{ bits/cycle})$
C	=	$T_a + T_d + 2T_P$
Q	=	$P - 80$

(b1)	C	=	$(89)/(10^6) + (256)/(10^6) + (2 \times 10^3)/(200 \times 10^6)$
		=	355×10^{-6}
	Q	=	176
	T	=	$(8 \times 10^6 \times 355 \times 10^{-6})/(176) = 16$ sec
(b2)	C	=	$(89)/(10 \times 10^6) + (256)/(10 \times 10^6) + (2 \times 10^3)/(200 \times 10^6)$
		=	44.5×10^{-6}
	Q	=	176
	T	=	$(8 \times 10^6 \times 44.5 \times 10^{-6})/(176)=2$ sec
(b3)	C	=	$(89)/(10^6) + (256)/(10^6) + (2 \times 10^4)/(200 \times 10^6)=445 \times 10^{-6}$
	Q	=	176
	T	=	$(8 \times 10^6 \times 445 \times 10^{-6})/(176)=20$ sec
(b4)	C	=	$(89)/(50 \times 10^6) + (10^4)/(50 \times 10^6) + (2 \times 10^3)/(200 \times 10^6) = 212 \times 10^{-6}$
	Q	=	9,912
	T	=	$(8 \times 10^6 \times 212 \times 10^{-6})/(9920) = 0.17$ seconds

(c) Note: We assume that in part (c) as in (b) there is an 80 bit header. We also use the same propagation speed as in (b).

Define T_r = Total repeater Delay = N/B. Again we assume an 80 bit header, and a propagation speed of 200 m/μs.

Then
$C = T_d + 2T_p + T_r$
$Q = P - 80$
$T = (8 \times 106 \times C)/(Q)$

We show results for N = 100

(c1) $C = (256)/(10^6) + (2 \times 10^3)/(200 \times 10^6) + (100)/(10^6)=366 \times 10^{-6}$
$T = (8 \times 10^6 \times 366 \times 10^{-6})/(176)= 16.6$ sec

(c2) $C = (256)/(10 \times 10^6) + (2 \times 10^3)/(200 \times 10^6) + (100)/(10 \times 10^6)$
$= 45.6 \times 10^{-6}$
$T = (8 \times 10^6 \times 45.6 \times 10^{-6})/(176)= 2.1$ sec

(c3) $C = (256)/(10^6) + (2 \times 10^4)/(200 \times 10^6) + (100)/(10^6)=456 \times 10^{-6}$
$T = (8 \times 10^6 \times 456 \times 10^{-6})/(176)= 20.7$ sec

(c4) $C = (10^4)/(50 \times 10^6) + (2 \times 10^3)/(200 \times 10^6) + (100)/(50 \times 10^6)$
$= 212 \times 10^{-6}$
$T = (8 \times 10^6 \times 212 \times 10^{-6})/(9912) = 0.17$ sec

3. Define:
PF_i = Pr[fail on attempt i]
P_i = Pr[fail on first (i- l) attempts, succeed on ith]

Then

$$E[retransmissions] = \sum_{i=1}^{\infty} iP_i$$

For two stations:

$PF_i = 1/2^k$ where k = MIN[i,10]

$$P_i = (\prod_{j=1}^{i-1} PF_j) \times (1 - PF_i)$$

$PF_i = 0.5$
$P_i = 0.5$
$PF_2 = 0.25$
$P_2 = 0.375$

$PF_3 = 0.125$
$P_3 = 0.109$
$PF_4 = 0.0625$
$P_4 = 0.015$
The remaining terms are negligible.

E[retransmissions] = 1.64

For three stations with K = MIN[i,10]

$$PF_i = 1 - \left(\frac{2^k - 1}{2^k}\right)\left(\frac{2^k - 2}{2^k}\right)$$

P_i is as before. A similar calculation leads to E[retransmissions] = 2.88.

CHAPTER 10

HIGH-SPEED AND WIRELESS LANs

ANSWERS TO QUESTIONS:

1. The speed and power of personal computers is growing explosively. Moreover, the MIS organization is recognizing the LAN as an essential computing platform, especially because of increased client-server computing and implementation of intranets. Both of these factors have dramatically increased the volume of data to be carried on LANs and reduced the acceptable delay on data transfers beyond the capability of 10 Mbps LANs.

2. Shielded twisted pair, high quality unshielded twisted pair (Category 5), relatively low quality unshielded twisted pair (Category 3) and optical fiber are transmission medium options for Fast Ethernet.

3. Fast Ethernet differs from 10BASE-T in two important ways. Two physical links are used between nodes -- one for transmitting, one for receiving (four links are used for Fast Ethernet using lower quality unshielded twisted pair (Category 3)). A different coding scheme 4B/5B-NRZI is used rather than Manchester Coding.

4. The Fibre Channel standard is organized into five levels:
   - FC-0: The Physical Interface and Media Level allows a variety of physical media and data rates.
   - FC-1: The Transmission Protocol Level define the signal encoding technique used for transmission and for synchronization across the point-to-point link.
   - FC-2: The Framing Protocol Level deals with the transmission of data between ports in the form of frames, similar to other data link control protocols.
   - FC-3: The Common Services Level provides a set of services that are common across multiple ports of a node.
   - FC-4: The Mapping Level defines the mapping of various channel and network protocols to FC-0 through FC-2 (the FC-PH).

5. FDDI and IEEE 802.5 are both token ring systems. The FDDI ring includes additional features designed to increase availability. For example, early token release is mandatory for FDDI and optional for IEEE 802.5.

CHAPTER 11

WIRELESS NETWORKS

ANSWERS TO QUESTIONS:

1. Universal personal telecommunications refers to the ability of a person to identify him or herself easily and use conveniently any communication system in a large area, e.g., globally, over a continent or in an entire country in terms of a single account. Universal communications access refers to the capability of using one's terminal in a wide variety of environments to connect to information services; e.g., to have a portable terminal that will work in the office, on the street, and on airplanes equally well.

2. The first generation systems of mobile telephones were based on analog communication using frequency modulation. The second generation uses digital techniques and time division multiple access (TDMA) or code division multiple access (CDMA). Advanced call processing features are present as well. The third generation will evolve from a number of second generation wireless systems. The idea will be to integrate these services into one set of standards. It is expected that a single handset will support a wide variety of services including voice, data, and video.

3. GSM differs from AMPS by having a Subscriber Identity Module which makes the subscriber portable rather than the terminal; it uses digital transmission with TDMA and FDMA rather than analog transmission. GSM communications are encrypted while AMPS are not. GSM was designed from the beginning to support data and image services.

4. You would use GEOs when the earth stations are not near the poles, when there is a premium on not having to steer the earth station antennas, and when broad earth coverage is important, for television broadcasting for instance. HEOs are primarily of use when coverage of areas near one of the poles is essential, such as the use of the Molniya satellites to cover the northern parts of the former Soviet Union. LEOs are useful for point to point communication, and for extensive frequency reuse. Since LEOs have much less propagation delay they are useful for interactive data services. They also can cover polar regions. Finally, while you need many more LEOs for broad coverage, each satellite is much less expensive than a GEO.

5. Frequency reuse can be achieved in wireless communications by using space division based on cells, and directional antennas. Code division can also be used to let multiple users use the same frequency.

6. LEO, GEO and HEO stand for low earth orbit, geostationary (or geosynchronous) orbit, and highly elliptical orbit, respectively. The traditional GEO satellite is in a circular orbit in an equatorial plane such that the satellite rotates about the earth at the same angular velocity that the earth spins on its axis. To accomplish this the satellite must be approximately 35,838 kms. above the earth's surface at the equator. LEO satellites are

satellites with much lower orbits, on the order of 700 to 1,400 kms. high. Finally, HEO satellites are characterized by an orbit which is an ellipse with one axis very substantially larger than the other. The height of the orbit can very, it is the **shape** of the orbit which characterizes this type of satellite.

Because of the high altitude of the GEO satellite the signal strength is relatively weak compared to LEOs. Frequency reuse is more difficult because the antenna beam (all other things being equal) covers a much greater area from a GEO than from a LEO. The propagation delay for a GEO satellite is about 1/4th of second; that of a LEO satellite is much less. Because the GEO satellite must be over the equator, the coverage near the north and south poles is inadequate. For these regions better communication can be achieved by LEO or HEO satellites. HEO satellites have the additional advantage that they spend most of their time at the "high" part of their orbit so that you get the most coverage for the longest time for this type of satellite. On the other hand, tracking and handoff is not necessary for GEO satellites because they appear stationary relative to the earth. LEO satellites, since they are so low travel very much faster, and cover less area than GEO so that tracking is more difficulty and passing off is frequent. HEOs require tracking and handoffs, as well. However, if the HEOs have high orbits the handoff frequency can be much less and the tracking easier than for LEOs.

CHAPTER 12

TCP/IP and OTHER PROTOCOL ARCHITECTURES

ANSWERS TO QUESTIONS:

1. The network access layer is concerned with the exchange of data between a computer and the network to which it is attached.

2. The transport layer is concerned with data reliability and correct sequencing.

3. A protocol is the set of rules or conventions governing the way in which two entities cooperate to exchange data.

4. A protocol data unit is the combination of data from the next higher communications layer and control information.

5. The data link layer is concerned primarily with adding flow control, error detection and control services to the physical link.

6. **Transmission Control Protocol/Internet Protocol (TCP/IP)** are two protocols originally designed to provide low level support for internetworking. The term is also used generically to refer to a more comprehensive collection of protocols developed by the U.S. Department of Defense and the Internet community.

7. Under IBM's System Network Architecture (SNA) network addressable units are the logical network components that provide communications interfaces through which end users can gain access to each other and through which the network is managed.

8. Under SNA, computers can achieve a peer-to-peer relationship using the APPC Standard.

9. A router is a device that operates at the Network layer of the OSI model to connect dissimilar networks.

ANSWERS TO PROBLEMS:

1. The OSI model could be further partitioned by adding a Media Access Layer to the Physical and Link layers. This would make the IEEE standards for LANs fit the OSI model better. Another possibility for an eighth layer could be an Internet Layer. On the other hand, the Session Layer could be absorbed in other layers without much loss to achieve a 6 layer model.

2. Data plus transport header plus internet header equals 1820 bits. This data is delivered in a sequence of packets, each of which contains 24 bits of network header and up to 776

bits of higher-layer headers and/or data. Three network packets are needed. Total bits delivered = 1820 + 3 x 24 = 1892 bits.

CHAPTER 13

DISTRIBUTED APPLICATIONS

ANSWERS TO QUESTIONS:

1. For a single system mail facility, terminal handling software is required as well as file handling software and a mail package that contains all the logic for providing mail related services to users.

2. The Simple Mail Transfer Protocol (SMTP) is the standard protocol in the TCP/IP protocol suite for transferring mail between hosts. Multipurpose Internet Mail Extension (MIME) is an extension of SMTP to address some of its problems and limitations. MIME addresses SMTP's inability to directly transmit executables and other binary files and provides support for national language characters that SMTP does not directly support.

3. Electronic data interchange (EDI) is the direct computer to computer exchange of information normally provided on standard business documents.

4. Some major benefits of electronic data interchange include speed, reduction of errors, greater security, and connection with other major automation applications.

5. An EDI system must (i) contain the application such as order entry, (ii) translation software, and (iii) a public or private communications network.

6. The three key elements of a Web system are the Uniform Resource Locator (URL), Hypertext Markup Language (HTML), and Hypertext Transfer Protocol (HTTP).

7. The URL consists of a name of the access scheme being used followed by a colon and then the identifier of a resource. Typical access schemes include "http," "ftp," and "gopher." The resource identifier is specific to the access scheme being used. Most important to the Web is the http access method.

8. The content of a Web page is specified using Hypertext Markup Language (HTML).

9. HTTP is a transaction-oriented client/server protocol. It makes uses of TCP. However, each transaction (e.g., page request) is independent. A separate TCP connection is set up for each transaction. Typically, the browser opens a TCP connection that is end-to-end between the browser and server. The browser then issues an HTTP request. When the server receives the request, it attempts to perform the requested action and, in any case, returns an HTTP response. The TCP connection is then closed.

10. For reliability, communication on the Web is based on TCP which, in turn, is based on IP.

ANSWERS TO PROBLEMS:

1. Mail-bagging economizes on data transmission time and costs. It also reduces the amount of temporary storage that each message transfer agent must have available to buffer messages in its possession. These factors can be very significant in electronic mail systems that process a large number of messages. Routing decisions may keep mail-bagging in mind. Implementing mail-bagging adds slightly to the complexity of the mail protocol and the message transfer agent software, but is usually worth it.

CHAPTER 14

CLIENT/SERVER and INTRANET COMPUTING

ANSWERS TO QUESTIONS:

1. **Client/Server computing** divides an application into tasks which are performed on the platform (computer) where it can be handled best. Clients request services from services which may be on the same or different platforms (computers). A single platform may be a client or a server, or both, for various applications and functions. That is, client/server computing is peer oriented rather than centralized or master/slave.

2. Client/Server computing is distinguished from other forms of distributed processing by (i) providing users with environments and resources at their workstations for user-friendly interaction with applications, (ii) centralizing of organizational databases and many network management and utility functions, (iii) emphasis on open and modular systems so that the user has a greater choice in selecting products and mixing different vendor's products, and (iv) an emphasis on networking to organize and operate the information system.

3. The user as client is usually provided with powerful presentation and graphical user interface facilities to provide the user individualized access to services. The servers provide access to shared facilities such as data bases, printing, and file storage. The actual application is divided between client and server in a way intended to optimize ease of use and performance.

4. "Fat client" refers to client/server configurations where a considerable fraction of the load is on the client. Conversely, "fat server" refers to client/server configurations in which the server carries the major part of the load.

5. The fat client approach takes advantage of desktop power, off-loading application processing from servers, making them more efficient and less likely to be bottlenecks. However, the addition of more functions to the client can overload the capacity of the client making upgrades necessary. Also, if the system supports many users, high capacity LANs may be needed to support the large transmission volumes between the thin servers and the fat clients. Finally, it is difficult to maintain, upgrade, or replace applications distributed across tens of hundreds of desktops. The fat server approach is useful for migrating from traditional mainframe environments.

6. **Middleware** is a collection of standard programming interfaces which provide a consistent and transparent interface between applications on the one hand and communications and operating systems on the other for a variety of server and workstation types in a client/server environment.

7. TCP/IP and OSI are communication standards; middleware provides facilities for distributed **computing**. Eventually, the TCP/IP and OSI standards may support the facilities to support distributed computing, but they do not exist today in these standards families.

8. Intranet is a term describing implementations of Internet technologies within a single organization, rather than for external connection to the global Internet.

9. An Intranet can be considered as a fat server realization of client/server technology. However, intranets are typically easier to deploy, use a small number of widely-accepted standards, integrate with other TCP/IP-based applications.

CHAPTER 15

DOING BUSINESS on the INTERNET

ANSWERS TO QUESTIONS:

1. The Internet grew out of the ARPANET.

2. Acceptable use policies define the rules of the road for using the network. For networks subsidized by public money, the acceptable use policies often limit commercial use. NSF's Acceptable Use Policy is one such.

3. The fastest growing Internet domain is ".com"

4. The most important barriers to the use of the Internet for on-line retail transactions are adequate communications, privacy and security, and payment systems.

5. SSL provides security, transparent to the user, at the lowest level of the protocol stack at a level just above the basic TCP/IP service. S-HTTP is an application level service.

6. The following features characterize an on-line payment system: (i) the type of transactions supported; (ii) the means of settlement; (iii) privacy and security aspects; (iv) by who takes on what risks; and (v) by the operational characteristics.

7. The classes of participants in SET transactions are: cardholders, merchants, financial institutions, and certificate authorities.

8. The most controversial aspect of ecash is its anonymity. Unless, the ecash user tries to double spend ecash, the transaction is not traceable. This raises concerns about untraceable money laundering, tax evasion, funding for terrorist activities, and other anti-social applications.

CHAPTER 16

NETWORK MANAGEMENT

ANSWERS TO QUESTIONS:

1. A fault is an abnormal condition which requires management attention (or action) to repair. A fault is usually indicated by a failure to operate correctly, or by excessive errors.

2. Fault tolerance is achieved by the use of redundant components and alternative routes.

3. The major categories of SNA network management are problem management, change management, configuration management, and performance and accounting management.

4. IBM views open network management as the ability of non-IBM suppliers to build NetView handles and extensions into their products based on IBM's own published set of interfaces and NetView operating details.

5. A non-IBM vendor can link his product to NetView for network management by interfacing to NetView/PC which has a documented software interface.

6. Network technical control includes using network performance statistics to facilitate planning by management.

7. A network control system provides centralized control and all technical control equipment activities. It gathers status and operational data from the equipment and responds to commands from the operator at the console.

8. A software monitor is a software package resident in main memory in a host or communications processor that can gather and report statistics on hardware and systems, and applications software activity.

9. The Simple Network Management Protocol (SNMP) is the network management tool for TCP/IP networks. It provides the following three basic capabilities: (i) *Get*: enables the management station to retrieve the value of objects at the agent; (ii) *Set*: enables the management station to set the value of objects at the agent; (iii) *Trap*: enables an agent to notify the management station of significant events.

10. SNMPv2 extends the original version SNMPv1 by providing support for decentralized network management, efficient transfer of large blocks of data, and, ultimately, security.

ANSWER TO PROBLEM:

1. (a) The value of a Gauge has its maximum value whenever the information being

modeled is greater than or equal to that maximum value; if the information being modeled subsequently decreases below the maximum value, the Gauge remains at the maximum value. The gauge can only be released by subsequent management action.

(b) The SNMPv2 interpretation provides a realistic representation of the underlying value at all times, subject to the limitation of the gauge. However, a manager may want to know that some maximum value has been reached or exceeded. By "sticking" the gauge at its maximum value until it is noticed and released by a manager, this information is preserved.

CHAPTER 17

NETWORK SECURITY

ANSWERS TO QUESTIONS:

1. Conventional encryption is based on a single key that is secretly shared between the parties involved in the information transfer. Public-key systems are based on two keys, one of which is public, and the other kept secret.

2. Data Encryption Standard (DES) is a conventional encryption scheme standardized by the National Bureau of Standards (now the National Institute of Standards and Technology) as a federal standard in 1977. It is the most widely used encryption scheme, but as the years have passed the capabilities of computers have increased to the point that the 56 bit key is becoming insufficient. In an effort to increase the effective key length, triple DES has been proposed which involves using DES three times in sequence with either two or three distinct keys.

3. It is important that the public key in a public key system be truly public in the sense that attacker can not substitute for the true public key, another key for which the attacker knows the corresponding private key. Rather elaborate key certification systems are set up to guarantee the authenticity of public keys.

4. To sign a document Bob calculates a hash function of the text to be signed. He then encrypts it with his private key, which presumably only he knows. This is the digital signature.

5. A digital signature is a cryptographic procedure which guarantees that a "signed" message is from the alleged source and has not been altered in any way after it was signed.

ANSWERS TO PROBLEMS:

1. The plaintext is "A good glass in the Bishop's hostel in the Devil's seat twenty-one degrees and thirteen minutes northeast and by north main branch seventh limb east side shoot from the left eye of the death's-head a bee-line from the tree through the shot fifty feet out" with spaces and punctuation added to clarify the text. For details see the "Gold Bug" by Edgar Allen Poe.

2. a) Each letter of the plaintext is mapped into a letter of ciphertext by the substitution obtained by taking the first letter of the quote, "t", as a plaintext letter and mapping it into the first letter of the alphabet "A". Similarly, each "h" in the plaintext is mapped into "B" in the ciphertext. This proceeds until it is time to assign the ciphertext letter "K" to a plaintext letter. The next letter in order is "t"

in "thick". But "t" has already been assigned. So has the next letter of "thick", "h". Finally, "K" is assigned to "i"; and so on. The assignment is indicated as follows:

```
ABC DEFG HIJ ABKLM FE ABC DACND IEO ABC DEFGPHIMCD OQKRCE SJ
the snow lay thick on the steps and the snowflakes driven by

ABC GKEO HFFMCO SHILM KE ABC BCIOHKTBAD FP ABC LIQD
the wind looked black in the headlights of the cars
```

To decipher the ciphertext the process is reversed:

```
SIDKHKDM AF HCRKIABIE SHIMC KD LFEAILA
basilisk to leviathan blake is contact
```

b) Such a scheme is reasonably secure if the book is secret. However, the substitution not being random makes it susceptible to frequency analysis. For example, letters that are frequent in the language of the plaintext; e.g., "e", "t", "a", "o", "n" in English will tend to receive cipher substitutes near the beginning of the alphabet.

c) It is better to use the first sentence of the book, rather than the last. If you use the last, there may not be enough text to provide a substitution for each letter of the alphabet. Starting with the first sentence you can continue until substitutions are obtained for all the letters of the alphabet.

3. "What do you make of it Holmes?"
"It is obviously an attempt to convey secret information":
"But what is the use of a cipher message without the cipher?"
"In this instance, none at all."
"Why do you say 'in this instance'?"
"Because there are many ciphers which I would read as easily as I do the apocrypha of the agony column: such crude devices amuse the intelligence without fatiguing it. But this is different. It is clearly a reference to the words in a page of some book. Until I am told which page and which book I am powerless."

Sir Arthur Conan Doyle, *The Tragedy of Birlstone*.